I0469282

Two Fatality Board and Care Facility Fire
Salvation Army Rehabilitation Center

Investigated by: Sheila-Faith Barry

This is Report 090 of the Major Fires Investigation Project conducted by Varley-Campbell and Associates, Inc. under contract EMW-94-C-4423 to the United States Fire Administration, Federal Emergency Management Agency.

 FEMA

Department of Homeland Security
United States Fire Administration
National Fire Data Center

U.S. Fire Administration Fire Investigations Program

The U.S. Fire Administration develops reports on selected major fires throughout the country. The fires usually involve multiple deaths or a large loss of property. But the primary criterion for deciding to do a report is whether it will result in significant "lessons learned." In some cases these lessons bring to light new knowledge about fire--the effect of building construction or contents, human behavior in fire, etc. In other cases, the lessons are not new but are serious enough to highlight once again, with yet another fire tragedy report. In some cases, special reports are developed to discuss events, drills, or new technologies which are of interest to the fire service.

The reports are sent to fire magazines and are distributed at National and Regional fire meetings. The International Association of Fire Chiefs assists the USFA in disseminating the findings throughout the fire service. On a continuing basis the reports are available on request from the USFA; announcements of their availability are published widely in fire journals and newsletters.

This body of work provides detailed information on the nature of the fire problem for policymakers who must decide on allocations of resources between fire and other pressing problems, and within the fire service to improve codes and code enforcement, training, public fire education, building technology, and other related areas.

The Fire Administration, which has no regulatory authority, sends an experienced fire investigator into a community after a major incident only after having conferred with the local fire authorities to insure that the assistance and presence of the USFA would be supportive and would in no way interfere with any review of the incident they are themselves conducting. The intent is not to arrive during the event or even immediately after, but rather after the dust settles, so that a complete and objective review of all the important aspects of the incident can be made. Local authorities review the USFA's report while it is in draft. The USFA investigator or team is available to local authorities should they wish to request technical assistance for their own investigation.

This report and its recommendations were developed by USFA staff and by Varley-Campbell & Associates, Inc., Miami and Chicago, its staff and consultants who are under contract to assist the Fire Administration in carrying out the Fire Reports Program.

The U.S. Fire Administration greatly appreciates the cooperation and information received from officials of the Miami Fire Department, most particularly Chief Carlos Gimenez, Deputy Assistant Chief Michael J. Essex, and Lieutenant Gene Cummings.

For additional copies of this report write to the U.S. Fire Administration, 16825 South Seton Avenue, Emmitsburg, Maryland 21727. The report is available on the Administration's Web site at http://www.usfa.dhs.gov/

U.S. Fire Administration

Mission Statement

As an entity of the Department of Homeland Security, the mission of the USFA is to reduce life and economic losses due to fire and related emergencies, through leadership, advocacy, coordination, and support. We serve the Nation independently, in coordination with other Federal agencies, and in partnership with fire protection and emergency service communities. With a commitment to excellence, we provide public education, training, technology, and data initiatives.

 FEMA

TABLE OF CONTENTS

Two Fatality Board and Care Facility Fire
Salvation Army Rehabilitation Center
Miami, Florida
November 11, 1995

Local Contacts: Chief Carlos Gimenez
Director of Department

Lieutenant Gene Cummings
Fire Investigator

Chief Michael J. Essex
Deputy Assistant Chief

Frank Barron
Assistant to the Fire Marshal

City of Miami Fire Rescue Department
Southwest 2nd Avenue, Tenth Floor
Miami, Florida 33130
305/416-1600

OVERVIEW

On November 12, 1995, a late night arson fire heavily damaged the second and third floors of a Salvation Army Adult Rehabilitation Center in Miami, Florida, and claimed the lives of two residents. Dozens of residents had to be rescued by firefighters using ground ladders, assisted by some of the early evacuees. Many of the 98 residents who escaped required treatment for smoke inhalation; three were transported in critical condition to local hospitals.

The first companies on the scene found residents hanging from the windows on the third floor, attempting to escape the heavy smoke. Some had knotted sheets into makeshift ladders and climbed out of the building, while most waited to be evacuated by fire department ladders.

Fire department personnel conducted interior search and rescue operations to evacuate the remaining residents, as well as providing medical assistance. Fire department rescue squads and private ambulances were besieged, as the victims sought medical assistance.

Investigators determined the area of origin to be a small open waiting area outside a group of offices on the second floor. Heavy smoke from the burning furnishings in that area spread throughout the second floor corridors and up to the third floor via open stairway doors. Most sleeping room doors

1

were closed, but the latching devices had been disabled. This prevented the doors from closing completely and contributed to the spread of the smoke.

The building was equipped with hurricane shutters which were installed on the outside of the building over the glass windows. The shutters hampered the residents attempting to exit through the windows, when the corridors became impassable because of heavy smoke conditions. Some residents used furniture to break out the shutters. There were instances of persons becoming stuck in windows from which the shutters had only been partially removed.

The building was equipped with a fire alarm system which consisted of pull stations and alarm bells in the corridors. Smoke detectors were installed in the sleeping rooms but were installed in only a few public areas. Smoke detectors were not connected to the fire alarm system. The fire alarm system had been deactivated by an on/off switch on the control panel.

KEY ISSUES

Issue	Comment
Fire Building	Three story masonry building used as residential treatment center.
Fatalities	Two residents died from smoke inhalation as they tried to exit smoke filled corridors.
Origin and Cause	Arson fire in an open waiting area adjacent to the second floor corridor. Combustible furnishings in the area of origin generated heavy smoke.
Fire Alarm System	The fire alarm system had been shut off.
Smoke Detectors	Smoke detectors had not been installed throughout the building.
Smoke Travel	Smoke spread quickly throughout the building through long undivided corridors, open stairway doors and partially open sleeping room doors.
Exterior Window Access	Heavy smoke in the corridors prevented residents from leaving their rooms. Hurricane shutters made it difficult for fire department personnel to evacuate residents with ladders.
Emergency Planning	Emergency planning or fire drills had not been conducted. Residents varied in their familiarity with building exits.
Triage response	Emergency personnel were overwhelmed by large numbers of victims.

THE BUILDING AND ITS OPERATIONS

The Complex

The Salvation Army Residential Center was home to adult men enrolled in a substance abuse treatment program. It is located between Miami Court and Northwest 1st Avenue near the downtown area. The Salvation Army complex contained the Residential Center, a Donation Center and a maintenance building.

The Donation Center is a warehouse facility where donations to the Salvation Army are received and processed. The building has several workshops for refurbishing donations and has complete automatic sprinkler coverage. A store is located on the first floor.

The adjacent maintenance building contained storage and supplies. The maintenance and warehouse buildings were not damaged as a result of the fire.

Fire Building Construction and Layout

The Residential Rehabilitation Center is located at 2236 North Miami Court. The building, which is three stories in height, was built in 1949 and is constructed with concrete block exterior walls, a steel and concrete structural frame and interior partitions. The second floor ceiling was constructed of drop-in ceiling panels on runners attached to twin-T concrete beams; on the third floor, the drop-in ceiling panels were attached to the bottom cord of the steel bar joists.

The building's main entrance opened directly onto the sidewalk on Miami Court. All other building exits opened within the confines of the fenced complex. Access to the rear of the building was through a chain link gate opening into a rear parking lot.

The first floor of the Center contained offices, a cafeteria and kitchen, a chapel and meeting rooms. The second floor contained additional offices, meeting rooms, store- rooms, sleeping rooms and several dormitory suites, which consisted of bedrooms arranged around a common living room.

An open air patio was located off the west side of the building at the second floor level. There was no outside exit to the ground floor from this patio. This was the only area in the building where smoking was permitted.

The third floor contained additional sleeping rooms and some dormitory suite arrangements. The floor was arranged with corridors surrounding a central core of rooms, including a communal bathroom which had open doors to corridors on both sides.

Most of the sleeping rooms held four beds; the number of occupants in each room varied. A few sleeping rooms were assigned to staff members or permanent residents, and had only a single occupant. There were some self-contained apartments with kitchen facilities on the third floor, and it also contained a TV room, a library, a barbershop and laundry facilities.

Emergency Exiting

The building had three enclosed stairways. The doors on the northeast (NE) and the southeast (SE) stairways, which were routinely used, had been propped open by the residents. The northwest (NW) stairway was seldom used, and the closed doors in this stairway provided a smoke free escape for many residents in that part of the building. Stairway doors were marked with illuminated exit signs.

Although equipped with self-closers, many of the sleeping room doors would not shut completely; many residents had tied strips of cloth across the lockplates to prevent the doors from latching when shut. The bulk of the cloth, which prevented the doors from closing completely, allowed smoke to seep around the door edges.

There were no smoke control separations constructed in the corridors. On the third floor, the corridor with the open central lavatory and the lounges were all effectively part of one large open space.

Each of the windows in the building was equipped with a permanent hurricane shutter. Installed on the outside of the operable glass windows, the shutters consisted of several interlocking metal louvers mounted on two vertical tracks. They could be opened and closed from the inside with a crank and lever mechanism, similar to awning windows. The shutters effectively precluded egress from the windows without their forcible removal. (See Appendix B, photograph 1.)

Furnishings and Interior Finish

Most of the furnishings in the building had been obtained over time from donations to the Salvation Army. There were a variety of furnishings in the public areas, mostly wooden and upholstered chairs, sofas, tables, magazine racks and bookcases, all of different age, manufacture and material.

Most of the sleeping rooms were sparsely furnished with single beds, wooden chairs and metal lockers. However, some of the single rooms, which were occupied by permanent residents, had been comfortably furnished by the occupants; some of these rooms contained bookcases, stereo cabinets, upholstered furniture and window draperies.

The office waiting area on the second floor where the fire originated was furnished with upholstered chairs, end tables and racks of magazines and promotional literature. Chair cushions were shown to be highly flammable in tests conducted by the fire department after the fire and shown to produce large quantities of thick black smoke.

Wallcoverings in the Center varied from plywood paneling and painted gypsum wallboard to vinyl wallcoverings. Investigators observed that the vinyl wallcovering in the area of origin, and leading down the second floor hallway from the area of origin, was completely burned away.

Fire Protection Systems

The building was equipped with a fire alarm system, consisting of audible devices and pull stations installed throughout. The system control panel was located in the watchroom on the first floor; there was no battery backup to the system.

Hardwired smoke detectors were installed sporadically throughout the corridors and other public areas. There were also smoke detectors installed in many of the sleeping rooms and dormitories. These detectors were not connected to the fire alarm control panel.

Pull stations were located near the stairway doors on each floor. However, the staff did not know if the pull stations were functional. No one reported attempting to activate the fire alarm system with a pull station during the fire. Investigators subsequently determined that none of the pull stations had been activated.

The fire alarm system control panel was mounted on a wall in a storage cabinet in the first floor watch room. The control panel was completely out of sight, hidden behind closed cabinet doors with cleaning supplies stacked in front of it. The Center staff, when questioned, was unaware of its existence. A switch of the type normally used as a wall switch had been installed in the control panel, providing an on/off switch for the system. When located after the fire, the switch was found in the "off" position.

There were no records or staff recollection to indicate when the fire alarm system had last been tested or inspected. The system was not connected to a central station or alarm monitoring service. There was no public announcement system in the building.

There were fire hose cabinets installed in the building corridors, and fire extinguishers were located on each floor near stairway doors. A partial sprinkler system supplied from the domestic water supply had been installed in the front lobby area; however, this area was not involved in the fire.

THE RESIDENTS

The Residential Center operates as a half-way house, providing counseling, lodging and other support to adult male residents with substance abuse problems. Each of the program participants is required to live in the Residential Center while enrolled in the treatment program, and is assigned a job at the adjacent Salvation Army Donation Center. The average stay is approximately one year, although some of the residents had been provided permanent staff positions and continued to live in the Residential Center after they completed their treatment.

The population of the Residential Center was fluid and varied from long-term residents who were familiar with the building layout to new residents who had limited awareness of stairways, exits and building configuration. The total number of residents changed often, as resident could enter the program at any time. The length of stay also varied by individual. At the time of the fire, 98 men were living in the building. All residents were able bodied individuals - none required nursing care or special assistance in exiting the building.

There was no accountability system for who was in the Residential Center at any particular time. A staff member was on duty 24 hours a day and a watchman was on duty in the office just inside the front entrance of the building (located at the southeast corner of the building). All outside doors were locked at 7 p.m.; access into the building after this time was through the front entrance, where the watchman would unlock the door. Residents were allowed to leave and enter the building at any time without checking in with a staff member. There was no sign-out book, which would have assisted in providing an indication of the building's daily census and resident location.

EMERGENCY PLANNING AND PREPAREDNESS

No fire drills had been held in the Residential Center for some time. Exit floor- plans were posted in the building, however, in at least one instance, the floorplan as illustrated on the posted exit plan was reversed. Some residents were not aware of the location of fire extinguishers or pull stations and neither were illustrated on the floorplans. Many residents were unaware of the existence of the third stairway in the northwest corner of the building. New resident orientation only briefly covered the subject of fire safety, and was mostly limited to restating the no-smoking policy. There was no written Emergency Plan.

NFPA 101, Life Safety Code, Chapter 311 calls for the preparation and periodic updating of an Emergency Response Plan for Board and Care facilities. The plan should outline the responsibilities of staff members in the event of a fire, and list the fire protection features necessary for the successful implementation of the plan.

NFPA 101 also requires at least 6 fire drills per year, to be held bimonthly. During each drill, residents and staff are expected to simulate an actual emergency situation, exiting the building through recognized fire exits. Routine fire drills in this building would have familiarized the residents with the NW stairway. In addition, simulating an emergency situation would have alerted the Center's staff to the condition of the disabled fire alarm system.

CODE RELATED ISSUES

The Center would be classified under NFPA 101 Life Safety Code as a Board and Care facility. With over 20 residents, it would be considered a large facility. An important consideration in determining the level of fire protection necessary for an individual board and care facility is the capability of the occupants to exit the building on their own, or with the assistance of attendants. The faster and more self reliant the population of a facility is, the less encompassing[1] the fire protection features must be.

Exiting capability is classified as "prompt", "slow" or "impractical". Prompt exiting capability indicates that residents would be expected to evacuate the building in similar time parameters as the general population.

The Life Safety Code identifies certain fire protection features that are required in a large Board and Care facility. These include a fire alarm and detection system with audible alarm signaling devices, smoke detectors in public areas, corridors and sleeping rooms, and manual pull stations. It would also require separation of sleeping rooms from corridors and public spaces.

Smoke barriers are required by the Life Safety Code between sleeping rooms and corridors and other common spaces. Doors from sleeping rooms to corridors must be equipped with self-closing devices. Such devices had been installed on the sleeping room doors in the Center, however, the effectiveness of the closing devices was negated by strips of cloth tied between door knobs across the lockplate which prevented the doors from latching.

The Center, with its population of able-bodied healthy adults, would most likely classify as a prompt evacuation capability, and as such, a sprinkler system would not be required. A partial sprinkler system had been installed in the first floor lobby, but did not discharge during the fire. It was not a factor due to the location of the fire.

THE FIRE

November 12 was a Sunday night. At 11 p.m. most of the residents were already in bed. "Lights out" in the Center was 10 p.m.--only the bathroom, hallway and TV room lights remained lit. Some residents were reading or watching television in the lounge areas; a few were standing in the hallway conversing as a group.

At approximately 10:30 p.m., a resident was ironing in the living room of a dormitory suite (see Appendix A) on the second floor. The door to the hallway was propped open. The resident noticed a glow from around the corner, in addition to some smoke. Going into the hallway to investigate, he turned the corner and saw flames in the office waiting area. He turned back and shouted a warning, then ran down the SE stairway and placed a call to 9-1-1 from the pay phone in the front hallway on the first floor.

Meanwhile, other residents on the second and third floors had noted the accumulating smoke, and were running from door to door, banging on doors and attempting to alert others of the fire. A few residents exited through the NE and SE stairways before the corridors became untenable. As heavy black smoke accumulated on the third floor, occupants were forced back into their rooms and started to break out windows.

One resident moved through the smoke from his room to the SE stairway, where a metal coat rack holding work shirts was located next to the stairway door. Touching the coat rack with his hand as he felt across the wall for the exit, he burned his hand on the coat rack. He managed to crawl back up the hall and reenter a room, where he was later rescued through the window.

---

[1]The City of Miami has adopted and inspects to the 1991 edition of NFPA 101, Life Safety Code.

Some of the men made their way down the NE stairway, and others used the SE stairway, exiting through choking smoke. A small group of men familiar with the seldom used NW stairway exited into the rear parking lot, without encountering smoke in the stairway.

Several of the residents exited on the second floor and congregated on the elevated open air patio located on the west side of the building. There was no direct exit from this area to the ground level. Some of the men dropped from the patio to the roof of the single story Salvation Army retail shop, then climbed down on a step ladder that was propped against the building by other occupants who had successfully exited via the stairways.

FIRE DEPARTMENT RESPONSE

The call to 9-1-1 was received through the Miami Police Communications Center at 10:44 p.m. A first alarm assignment consisting of Engine 2, Engine 5, Aerial 5, Rescue 2, Squad 2 and a District Chief was dispatched at 10:47 p.m. The first unit on the scene, Engine 2, arrived at 10:49 p.m. and observed smoke from the third floor windows. Residents were breaking out windows in an attempt to escape the smoky conditions.

Interpreting the smoke as a mattress fire on the third floor, the crew of Engine 2 pulled a 1-3/4-inch pre-connected line from the engine and proceeded up the SE stairs, past the second floor, stopping midway between the second and third floors. Heavy heat and smoke at this level pushed them back down to the second floor landing, where they observed a glow from the office waiting area. They opened the hoseline into the waiting area and quickly controlled the visible flames.

Meanwhile, Squad 2 had arrived and pulled a second pre-connected 1-3/4-inch line to the second floor office area, backing up Engine 2. They moved through the second floor, knocking down spot fires and searching for victims.[2]

Although the fire was quickly controlled, the heavy smoke and heat had filled the corridors and many of the rooms, trapping residents on the second and third floors.

Exiting the building to replace their air cylinders, the crew of Squad 2 found the first victim laying face down on the sidewalk, apparently having suffered a cardiac arrest. Treatment was initiated by the crew of Squad 2; the victim was then transported by Rescue 2. This casualty was treated and released.

Center staff and residents crowded the area under the windows, attempting to help fellow residents escape by using step ladders available on the premises. One resourceful individual mounted a step ladder in the bed of a pickup truck and drove under windows in the rear parking area to allow residents to reach the ladder.

Pulling a third line from Engine 2, the crew of Aerial 5 proceeded to the third floor directly above the office waiting area, and checked for extension of fire in the wall space. No extension of flame was found. They opened a roof panel above the third floor stair landing for ventilation.

After meeting up with Engine 2 at the east side of the building, the crew of Engine 5 raised a 30 foot extension ladder and a 50 foot Bangor ladder to assist residents from the third floor windows.

---

[2] Miami Fire Department squads are composed of one firefighter/paramedic and one firefighter/emergency medical technician on an ALS truck which is equipped with rescue and basic firefighting gear. Squads respond on both EMS and firefighting calls, and are used for BLS transport and to provide additional manpower on a fire assignment.

The District Chief arrived at 10:50 p.m. finding heavy smoke issuing from the second and third floor windows and multiple victims trapped on the third floor, he called for a second alarm at 10:55 p.m. He instituted the Incident Command System, and established search, rear and medical divisions. He also declared a Mass Casualty Incident, which automatically calls for three additional rescue trucks, two suppression units and three squads, in addition to notification of the department director, the fire marshal and the local trauma unit medical staff.

Engine 12, Engine 6, Aerial 1 and Engine 4 responded on the second alarm. The crews of Engine 12, Engine 4 and Aerial 1 were assigned to search and rescue. The crew of Engine 6 evacuated third floor residents by means of ground ladders.

Search and rescue efforts on the second and third floors continued. The crew of Engine 12 crew located two unconscious victims in the hallway outside the bathroom on the third floor. They were removed and transported to a local hospital, where they were pronounced dead.

Arriving at 10:49 p.m., Engine 5 laid a 5 inch supply line from the hydrant at Miami Court and 23rd Street to Engine 2.

While conducting a door to door search on the third floor, the crew of Aerial 1 located an occupant trapped in one of the sleeping rooms. Using buddy breathing they moved the occupant to the first floor. He was transported to the hospital and later released after treatment for smoke inhalation.

As initial search and rescue efforts were completed, smoke ejectors were set up on the second floor to help clear the heavy smoke from the building.

TRIAGE EFFORTS

Ten private ambulances and eleven Fire Department rescues had been called to the scene, and were soon busy assisting the evacuees. Although many were treated on site with oxygen for smoke inhalation, most refused to be transported to a hospital. One firefighter incurred a minor facial burn injury and was treated and later released.

As residents exited the building, they converged on the triage area, seeking treatment. The demands of the victims overwhelmed the rescue crews, who provided immediate attention to those victims requiring critical care. Many of the residents waited in the immediate area for assistance, crowding the treatment area.

Employing the Med-Tag Triage system, paramedics attached colored triage tags to patients, designating the priority of treatment. Red tags identify critical patients who require immediate attention; yellow tags indicate patients who require attention within an hour. Green tags are used for lower priority patients who do not have life threatening injuries (often referred to as the "walking wounded"). Many of the "green tag" victims went to the treatment area when directed to do so, but lost patience and walked away from the area when they were not immediately attended to.

At the start of the incident ten private ambulances were assembled in a staging area for use as transport vehicles. During the incident, some members of the ambulance crews left their vehicles and joined the evacuees who were wandering around the fire scene, rendering their ambulances unavailable for transport. Sixteen victims were transported to the local trauma unit, most of them in Miami Fire Department rescue and squad trucks. This overtaxed fire department EMS personnel by decreasing the number of available units and personnel at the scene. The trauma center was nearby and turnaround time was short.

AFTER THE FIRE

Of the sixteen victims who were transported to area hospitals, six were treated and released, eight were admitted, three to critical care units--including the cardiac victim, and two were DOA. Autopsy results indicated smoke inhalation as cause of death. Most of the injuries were related to smoke inhalation, but one resident was admitted for treatment of a broken leg incurred when he jumped from a third story window.

Forty-eight units were on the fireground when the incident was brought under control at approximately midnight.

Investigators from the City of Miami Fire Department, Arson Division, established the area of origin as the office waiting area on the second floor and concluded that the fire had been deliberately set. A discharged fire extinguisher was found in the area of origin, however, it was not determined who had used it.

ANALYSIS

The actual fire was contained to a small area, and quickly controlled by the first arriving engine company, however, the unimpeded spread of heavy smoke compromised two floors of this large building. Only a few moments after the first recognition of a problem, the hallways on both the second and third floors were filling with thick black smoke, forcing residents back into their rooms. A functioning fire detection and alarm system equipped with the code recommended number of smoke detectors and alerting devices would have provided earlier warning of the developing fire. An earlier warning of the fire would have allowed time for the residents to evacuate before dangerous levels of smoke accumulated in the hallways.

There was no record of inspection or testing having been performed on the fire alarm system. The staff was unable to identify or locate the fire alarm control panel. The inclusion of the illegal on/off switch effectively disabled the fire alarm system but fire department investigators were not able to determine if the system was operable.

One fortunate characteristic of this fire scenario compared with other Board and Care fires was the general health and alertness of the resident population. The residents were healthy adults, without serious medical, mobility or mental awareness limitations, and capable of self-evacuation. When confronted with smoke filled hallways, most of the residents had the physical capability to break out the shuttered windows and climb unassisted down makeshift escapes and fire department ladders.

The lack of smoke separations in the corridors contributed to the rapid spread of smoke throughout the second and third floors. Stairway doors were propped open and provided vertical routes for the smoke to spread. As smoke filled two of the three stairways, those emergency exits were effectively nullified, forcing residents to depend on egress through the shuttered windows for escape.

LESSONS LEARNED

1. **Adequate functioning fire detection and alarm systems are critical.**

    A working fire detection and alarm system would have alerted the staff and residents to the fire's danger and the need to evacuate. The early warning factor is particularly important in a facility with fluctuating and sometimes transient populations, where evacuation time may be longer because occupants are unfamiliar with the building layout. A working fire detection and alarm system warns residents of a fire; it is no longer necessary to rely on staff and fellow residents to alert others individually of a fire. Many of the residents had already retired for the evening; an alarm system would have provided additional time to escape.

2. **Emergency plans and fire drills identify problems, increase awareness and are integral to the fire safety of a facility.**

    Both staff and residents of Board and Care facilities should be aware of and trained in what to do in an emergency situation. An up-to-date Emergency Plan and routine fire drills are essential to provide this awareness. Regular drills can highlight shortcomings or problems with the Plan or with the facility's fire protection systems. For example, a fire drill held in the Salvation Army Center would have revealed that the fire alarm system had been shut off. Drills also increase general awareness in the resident population, and may improve reaction time to a fire situation.

3. **Routine and thorough inspections by the fire department for code violations and unsafe conditions can prevent tragedies.**

    The facility had not been recently inspected by the city Fire Marshal's office and there were no indications that the fire alarm system had been tested or inspected by a fire alarm service. Routine inspections could have provided the opportunity to address the shortcomings of the fire alarm system, in addition to assisting the Center's staff with emergency planning, fire drills and other educational services to improve the Center's fire safety.

4. **Incidents with large numbers of casualties require procedures to effectively identify, process and treat victims.**

    Effective planning and practiced procedures are necessary to deal with mass casualty incidents. A good plan should stress organizing patients by medical need as well as decreasing congestion at the treatment sites. Although the majority of the patients in this fire were adequately treated at the scene, they crowded the treatment areas and wandered throughout the area. Some were left untreated after seeking attention for minor injuries.

    There should be a procedure to track ambulatory patients in the treatment area to ensure that follow-up treatment is administered, especially in cases where more critical patients delay treatment of large numbers who are less seriously injured. Holding areas must be monitored by individuals who can provide assistance until a secondary exam is performed and the patients are released. The police department can provide security and control of the treatment area.

5. **All responders in an incident should be aware of their assignment and be available to perform it when needed.**

    Ambulance drivers who left their vehicles made them unavailable for patient transport. This taxed the Fire/Rescue resources on scene, as they had to provide both treatment and transportation. The importance of fulfilling assignments should be stressed to all participants in an incident, including individuals who may be unfamiliar with structured operational plans, such as mutual aid personnel. Response personnel must be assigned to direct mutual aid resources who may not be familiar with the operational plan. Mass Casualty Drills involving the fire department and other service providers can be an effective means to familiarize participants with operational plans and task assignments.

APPENDIX A

General Site Arrangement

Floor Plan: 2nd Floor

Floor Plan: 3rd Floor

Appendix A (continued)

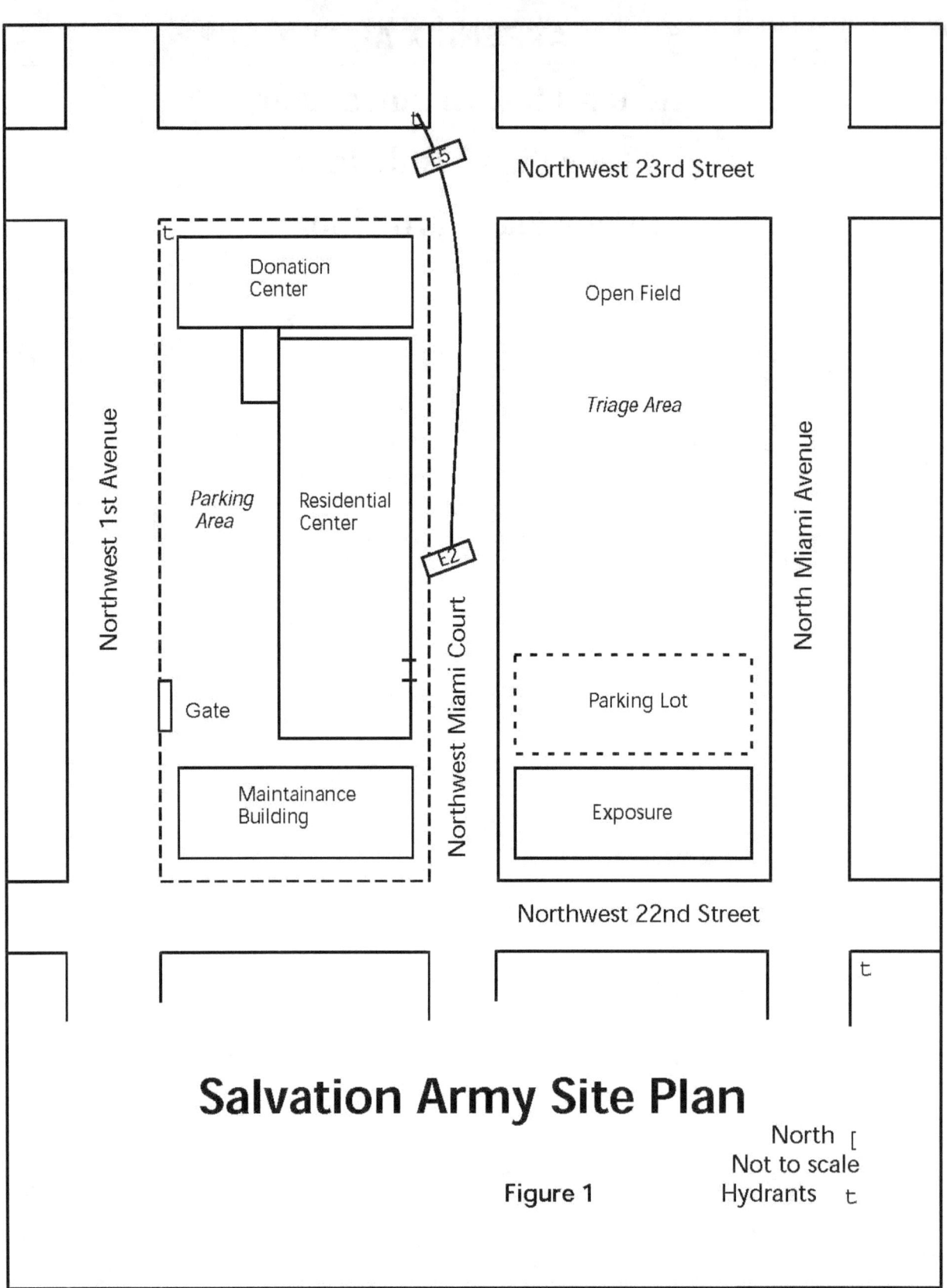

Salvation Army Site Plan

North

Not to scale

Figure 1   Hydrants   t

Appendix A (continued)

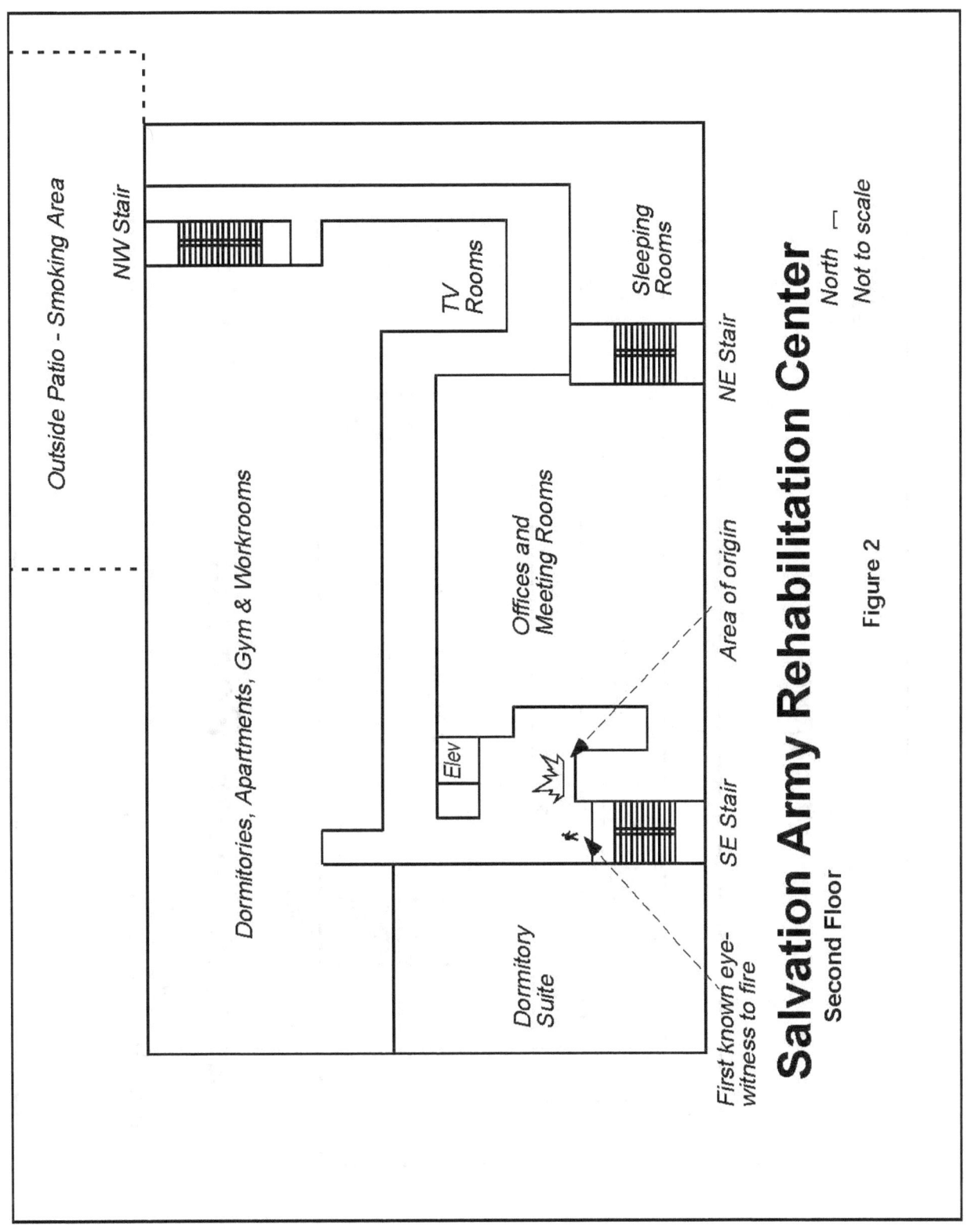

Salvation Army Rehabilitation Center
Second Floor

Figure 2

Appendix A (continued)

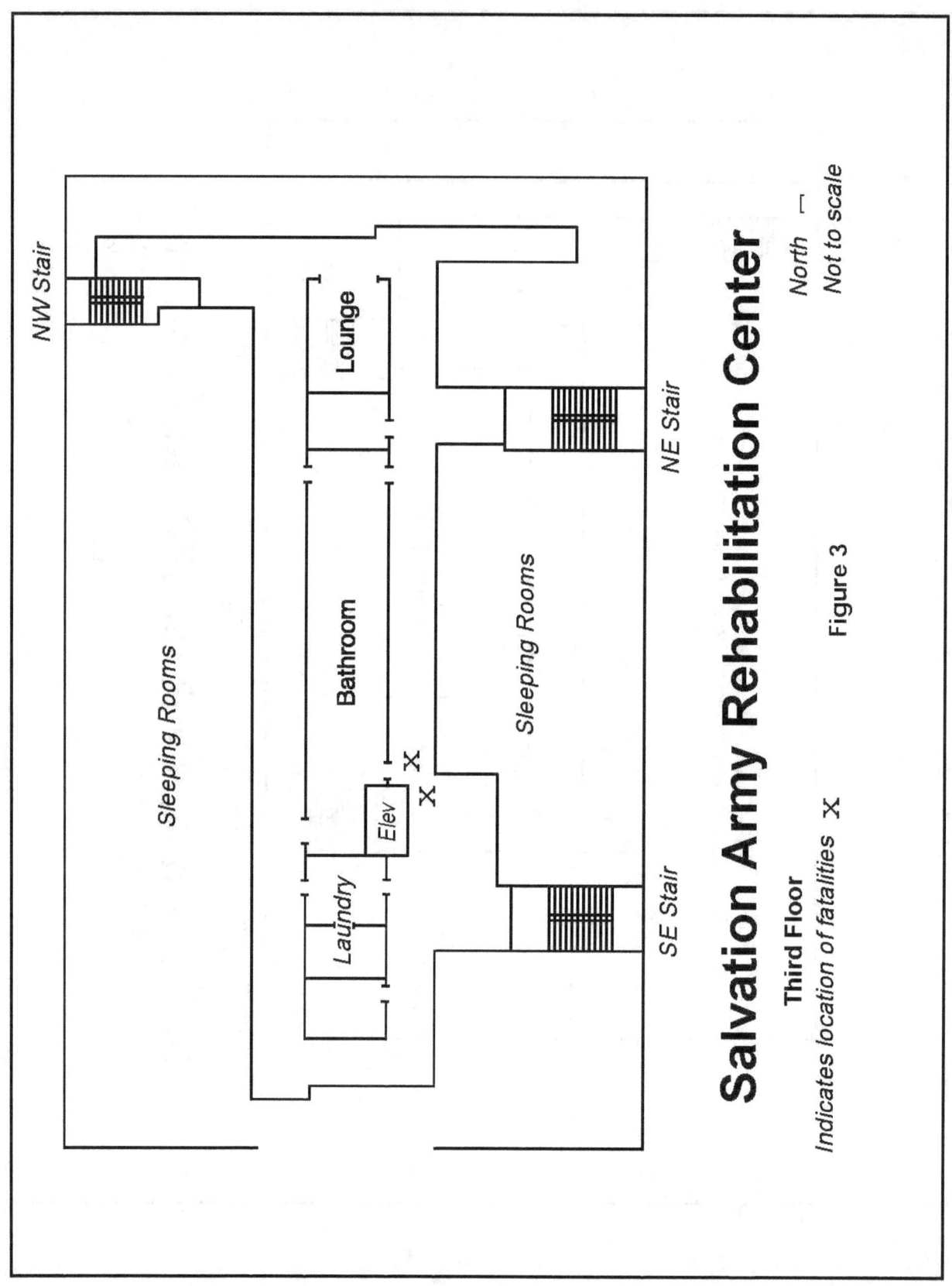

Salvation Army Rehabilitation Center

Third Floor

Indicates location of fatalities X

Figure 3

North

Not to scale

NW Stair

NE Stair

SE Stair

Lounge

Bathroom

Sleeping Rooms

Sleeping Rooms

Laundry

Elev

APPENDIX B

Photographs

Photographs were taken by the Miami Fire Rescue Department

1.   Southeast corner of the Rehabilitation Center. Front entrance is just north of the razor wire topped wall. Note permanent hurricane shutters installed on the outside of the windows. Some residents tied sheets together and climbed from the windows after breaking out the shutters.

Appendix B (continued)

2.   Doorway opening from SE stairway into office waiting area on second floor, adjacent to where investigators determined the fire originated.

Appendix B (continued)

3.   View of office waiting area on second floor from the SE stairway door.

Appendix B (continued)

4.   First view of hallway leading from office waiting area to residential areas of the second floor.

Appendix B (continued)

5.   Second view of hallway leading from office waiting area to residential areas of the second floor.

Appendix B (continued)

6.   Fire alarm control panel. The panel was located inside a cabinet in the first floor front office. Note the on/off switch in the bottom left hand corner of the box.